Robert P. Traxler

**The Principles of Mechanics as Applied to the Solar System**

Robert P. Traxler

**The Principles of Mechanics as Applied to the Solar System**

ISBN/EAN: 9783744749879

Printed in Europe, USA, Canada, Australia, Japan

Cover: Foto ©berggeist007 / pixelio.de

More available books at **www.hansebooks.com**

# THE

# Principles of Mechanics

## AS APPLIED TO

## THE SOLAR SYSTEM

#### WITH

ILLUSTRATIONS, SHOWING BY RADIATING LINES THE
MANNER IN WHICH THE FORCES OF THE SUN ARE
APPLIED TO THE PLANETS, AND THE MAN-
NER IN WHICH THE FORCES OF THE
SUN AND PLANETS EMANATE
FROM THEMSELVES.

#### ALSO,

THE CAUSES OF MAGNETIC CURRENTS, HEAT, OCEAN
CURRENTS, EARTHQUAKES, ETC.

#### AND THE

PRINCIPLE OR CAUSE OF THE TIDAL ACTION
ILLUSTRATED.

SAN FRANCISCO:
C. A. MURDOCK & Co., PRINTERS.
1889.

# PREFACE.

As the science of Astronomy has, from time almost immemorial to the present, been compelled to carry with it theories which have been and are now more or less speculative in their nature and character, the author hopes, in placing this little work before the reader, that it will not be felt as an additional burden, but will be kindly accepted and considered, and that the theories that it contains will be carefully compared with all applicable natural phenomena and principles, in mechanics, with which the reader may be familiar and that the claims that are herein advocated may be sustained only by the merits which they possess.

Most of the illustrations which are used as explanatory of the substance matter contained herein are necessarily exaggerated in regard to sizes and distances, etc., the same as nearly all astronomical illustrations, and they are used mainly to show the application of the theory or principle involved.

The author is well aware that the apparent movements of the satellites of the two farthest planets, also those of some of the comets may be referred to as exceptions, but as there are evidences of forces in operation at those remote distances, the same as between those bodies and the Sun, it is the opinion of the author that it is only necessary to wait for greater scientific advantages, which will enable us to investigate further and more carefully, when natural forces will be found to be operating upon them in a strictly mechanical way.

The diameters, distances, times of revolution, etc., used by the author are from the more recent calculations.

It has been the effort of the author to describe and illustrate the claims herein set forth by principles that the general reader can readily understand and with which the common experiences of life familiarize us. The use of technical terms has been carefully avoided, as much as possible, so that the reader, casual or otherwise, may be better able to reject or approve of the idea presented to the mind for consideration.

It has likewise been the aim of the author to reject the many opportunities to incite the mind of the reader to wonder, astonishment or admiration, preferring rather to represent as nearly as possible the operations of our planetary system within a space that will enable the mind to comprehend the movements of the planets and comets revolving around the Sun, making the solar system appear as a simple and natural combined piece of mechanism, or a mere toy of the universe.

Finally, in committing to the reader the few thoughts which are contained in this little work, it is the belief of the author that the more the divine laws, in all things, are studied and the better they are understood, the more munificent will they appear.

# CONTENTS.

# ILLUSTRATIONS.

# PRINCIPLES OF MECHANICS AS APPLIED TO THE SOLAR SYSTEM.

The comparative sizes of the Sun, Jupiter, the Earth and the Moon, as shown in Figure I, are intended to give an idea of the power and force of the Sun as acting upon the planets from the center of the solar system. The proportional size of the Moon is shown only where the figures of Jupiter and the Earth are enlarged. The diameter of the Sun, as shown, is sufficiently large to include about ten diameters of Jupiter when placed in a straight line, and the diameter of Jupiter equals about eleven diameters of the Earth, while the Earth's diameter embraces nearly four diameters of the Moon, the diameter of which is about two thousand one hundred and sixty miles. The diameter of the Moon compares well with the distance from San Francisco to Chicago, or from San Francisco to the Sandwich Islands, or to the length of the Mediterranean Sea.

## THEORIES OF PLANETARY MOTION.

Most of the theories advanced by astronomers, within the last two or three centuries, in regard to planetary motion, position, etc., have been accepted and readvocated by nearly all who have followed them in the study of that science down to the present time, with but few dissenting statements.

Ptolemy guided the greatest researchers in astronomical science into error for nearly thirteen hundred years; and although Kepler was one of the principal men to break through the barriers of erroneous traditions and beliefs, yet he retained a belief in the music of the spheres, which is now conceded to be one of the greatest fallacies which was

attached to the astronomical science. Yet all that he gave
us that is true is just as valuable as if he had never advocated theories which have since been discarded. From before his time to the present, many of the theories have been,
are still, and no doubt will remain, speculative.

Theories which have been advanced to account for the
various movements of the planets and planetary substances
have not been accepted as harmonizing entirely with natural forces, and so the cause of their movements has been
only partially explained. The law of attraction and gravitation, as set forth by authors generally, is almost univers-
ally accepted as being correct; but a plausible theory showing the development of the repelling force and its manner
of counter-balancing the planetary bodies and of rotating
them on their axes and around the Sun, also the satellites
around their primaries, etc., has heretofore ·been discourag-
ingly sought for by many earnest searchers in the science
pertaining to heavenly bodies.

The author attempts herein to show the manner and the
harmony in which the two forces are acting together upon
the planetary bodies and substances where one force is
always trying to draw them to itself, while the other is
always keeping them certain distances away.

The repelling force, as herein set forth and illustrated, is
claimed to be produced by the centrifugal force which is
developed by the rotation of the Sun and planets on their
axes, and which is thrown from the equatorial surfaces of
the revolving bodies in sufficient amount to keep the planets
and bodies upon which it is acting at the various distances
at which they are seen, and also to rotate them on their
axes and revolve them in their orbits. This same force is
shown to be operating upon the more distant planets in the
same manner, and these are in each case drawn by the ever-
acting law of attraction, and always met by the repelling
force which causes their varying movements according to
the amount and manner in which this force is applied.

In mechanics, estimates of power required to accomplish
stated results can be made with tolerable accuracy, with
margins for loss by radiation, friction, etc., which must

all be considered in the economic result, or otherwise, of the object to be attained.; but in the solar system, there are no indications as to the economy of forces, either of attraction or of repulsion.

The power of attraction appears to be a natural and an established force pervading all space and the principal agency by which the substances of the celestial centers of the universe have been drawn together and by which has been effected the building up and sustaining in form of the planetary and gaseous bodies of our solar system.

The repelling action of the Sun and of the revolving planetary bodies is also a law natural to them, for by their rotation they are always throwing off the repelling force, whether a planetary body or substance upon which it can act be within range or not.

### THE SUN.

The Sun is the great central body of attraction, light and force of this solar system, towards which all of the planets and planetary substances are drawn, and from which they are all repelled, and around which they all revolve. It attracts or draws to itself, in direct lines, by the law of attraction, from every direction in space around it, from the north and south as well as in the direction in which the planets are found. Its light is presumed to radiate from every part of its surface in direct lines and in equal quantity and brightness on its northern and southern surfaces, if not more so than on or near the equator.

The Sun is about 866,000 miles in diameter, and rotates on its axis, from west to east, once in about twenty-five days. It seems to be surrounded by a gaseous envelope or influence which appears to adhere to it. The depth of this envelope is not known, but it is probably extensive, as eruptions have been observed on its surface which extended to a height of more than 200,000 miles without being disturbed by any external influence, thereby showing that at that distance from the Sun the gaseous envelope revolves with it. This gaseous influence which adheres to and revolves with the Sun is probably of greater depth in the region of the Sun's equator than on its northerly or southerly surface, for it would be drawn away from the region of the poles and towards the equator by the rotation of the Sun on its axis, as substances or influences always seek the farthest point from the axis of a revolving body before being thrown off.

The development of the repelling forces is caused by the rapid rotation of the Sun on its axis. These influences or forces which are thrown off seem to be drawn to the Sun, from the space at each side of its equatorial region, and carried by the Sun's rotation to or near the extreme edge of the gaseous envelope, at or near the Sun's equatorial plane, there to be thrown off as the repelling force, while new influences or elements are continually being drawn to the Sun and replacing them only to be thrown off in the same manner. This repelling force seems not only to be thrown off from the extreme edge of the gaseous envelope, but some of it appears to be developed and thrown out from all parts or distances between the outer edge of it and the surface of the revolving body, and it seems to permeate and operate through any influences which may be outside of it.

Whatever this repelling force is, and whether acting singly or combined, whether electrical or otherwise, is not yet known; but the disturbances of the magnetic needle wherever these forces are interfered with favor the theory of electric repulsion; nor is it yet known that there are not natural agents, yet undiscovered, that will equal if not excel it in subtlety, wonder and usefulness.

There seems to be nothing surrounding the Sun or any of the planetary bodies to indicate a throwing-off force, except in the regions of their equatorial planes.

Many of the photographs of the Sun which have been taken during the time of total eclipse show streams apparently leaving it in greater amount and reaching farther from its surface in the region of the equator than elsewhere.

The principle that keeps a ball or sphere suspended and revolving in a jet of air or water when emitted vertically with sufficient force to overcome the attraction of the Earth is similar in action to the force that is constantly and simultaneously thrown off from the entire circumference of the Sun and all of the rotating planetary bodies or any rotating sphere at or near their equators.

The principles of mechanics as herein applied and illustrated are intended to refer, in general, to the manner in which the repelling force is set in motion and applied to the planetary bodies, and to the results as shown in the planetary movements. It is herein claimed that all of the planetary bodies, while revolving around the Sun and rotating on their axes, are only acting in harmony with and obedient to a superior force which is naturally and mechanically acquired and applied to them.

In applied mechanics it is well known that if the rotation or motion of a body which produces or supplies power is maintained at a regular and uniform speed, a steady and uniform result will be produced wherever this power is so applied; but when the force of a power is varyingly applied, corresponding results are expected. So, the author claims, in regard to the planetary motions, that wherever varying motions of the planetary bodies are seen, we may expect to find the force that keeps them in their orbits applied to them in a correspondingly varying manner also.

As will appear farther on, and as shown in the sectional edge-view of the Sun, in Figure II, this force of the Sun is not equally applied to the planetary bodies during their entire orbital circuits, for all planets cross this force at a small angle to it twice in each of their revolutions around the Sun, and the superior amount of force which is applied to them at times when crossing it results in their varying orbital distances; but the rotation of the Sun and of the planetary bodies on their axes is probably uniform with most of them, for bodies as large, cumbersome and heavy as any of these must have gained by their rapid rotations so great a momentum that other than a uniform axial rotation must be an unnatural movement.

The axes of the Sun and of each of the rotating planets are maintained in their same general directions by the revolution of the planetary substance matter around their axial centers. These substances are moving in direct or straight lines, excepting the regular curve which they make around their centers, according to their radial distances. The velocity and momentum of the planetary substances always depend on the distances that they are from the axis of revolution. As the entire amount of the Earth's substance revolves around its axial center once in about twenty-four hours, the velocities of this substance vary all the way from nothing, at the axis, to more than one thousand miles per hour, on the surface at the equator. This would indicate an average speed and momentum which would about equal that of the Earth, if it were moving in a straight line with the velocity of three or four hundred miles per hour. The amount of power required to cause any degree of lateral movement of a body of like dimensions, if going at a velocity of three or four hundred miles per hour, should indicate a proportional power that would be required to change the plane of the Earth's equator in a corresponding degree. This same principle operates in maintaining the direction of the axes of all of the revolving planets, the axis of the Sun and the axes of all substances whatsoever which have a rotary motion, even to the child's spinning-top, except that the top is influenced more by friction and attraction, which operate against its movements. The top is sustained in its tilted position by the velocity at which one-half of its volume is being constantly moved contrary to the attraction of the Earth and with a speed and momentum which exceed the Earth's attraction for it, and as there are only the centrifugal force of the top and the attraction of the Earth as the principal forces in operation connected with it, it may revolve in a tilted position until its speed decreases to a degree at which the attraction of the Earth exceeds the centrifugal force of the top, at which time it will suddenly fall. The natural tendency of the top, while in motion, is to gradually assume a vertical axial position. The principle of maintaining a general axial direction is also shown by the fact that a heavy top will resist considerable force, when rotating rapidly, if an effort be made to quickly change its plane of rotation. (See *Scientific American*, October 9th, 1886, page 230.)

FIG II

A SECTIONAL SIDE AND EDGE VIEW OF THE SUN
WITH MERCURY VENUS EARTH AND MARS

The Sun and each revolving planet are developing and throwing off this repelling force or influence at the region of their equators and in the direction of their equatorial planes, and in amount, according to the influence which surrounds them, with a steady and uniform current, and with a force according to their individual axial velocities. This repelling force, put in motion by the Sun, is presumed to extend from the Sun far into the space beyond the planet Neptune, while the force developed by the planets extends from them considerably beyond the orbits of their most distant satellites, as shown in Figure IV.

It is intended to illustrate in the sectional side and edge-view of a part of the planetary system, as shown in Figure II, the manner in which the Sun's force is distributed by its rotation and the manner in which the influences approach the Sun; also the positions of some of the planets as being acted upon and controlled by the Sun's force. In the sectional edge-view, the poles of some of the planets are shown in positions similar to those of the Earth at the time of the Summer and Winter solstices, while the equatorial plane of the Sun is represented by a dotted line in position similar to its relative angle to the ecliptic at its equinoxes. The angles of all of the orbits are shown as varying in degree from the Sun's equatorial plane. The ecliptic, or the plane or level of the Earth's orbit, is shown by being extended with dotted lines. From this plane, astronomers take measurements in the heavens. In other respects, it has no more significance than the orbit of any other planet, as nearly all of the varying movements of the planets depend on the extent of the Sun's force as applied to them. If the planets were observed and studied as intersecting the plane of the Sun's force, also the satellites as intersecting the equatorial planes of their primaries, much information might be gained that would lead to a knowledge of planetary action that cannot be so easily obtained by only a knowledge of the intersection of them with the ecliptic.

In the sectional side-views, are seen the Sun and planets in their respective positions, and all rotating on their axes, from west to east, the planets traveling around the Sun in the same direction. In the same views are shown dotted lines representing the direction of the attraction of the Sun for the planets, as extending from the center of each planet to the center of the Sun, and lines representing repulsion extending in tangents from the Sun to the planets. By the aid of this illustration, it can be seen that much more force is being applied to one and the same side of the line of attraction of all or each of the planets than to the other side. If this repelling force is powerful enough to keep them all at their respective orbital distances, then it seems evident that it is also powerful enough to rotate them all on their axes in the direction of least resistance and in their orbits in the same direction.

## MOTIONS OF THE SUN.

Spots have been observed, apparently crossing the Sun's surface, by which the time of its axial rotation has been ascertained. At about the time of the Winter solstice, or

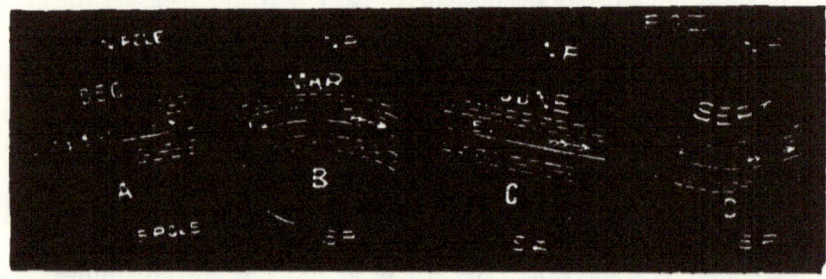

near the end of the year, the spots appear to travel straight across the Sun, ascending a little as they advance, as shown in the diagram at "A," Figure III. When observed toward the last of March, or about the time of the Vernal equinox. the spots appear to travel across the Sun in a somewhat bowed path, causing them to appear to rise and fall a little while crossing the Sun as shown at "B." At the time of the Summer solstice, or near the last of June, they are seen to travel straight across the Sun again, descending a little as they advance, as shown at "C." When observed from the direction of the Autumnal equinox, or near the last of September, they appear to descend and rise a little as they advance, as shown at "B." The changes which appear to occur in the paths of the spots in crossing the Sun are produced gradually, only reaching the extremes at about the times stated above, and, if observed at any intermediate time, they would appear dipped or bowed in proportion to and according to the time in which they are observed. These paths show the equatorial plane of the Sun to be at an angle of about seven degrees to the ecliptic. The spots appear more abundantly in periods of about eleven years, while at intermediate periods they sometimes almost, if not quite, disappear. Sometimes some of the spots are of short duration, often less than half an hour, while some do not wholly disappear until six or seven months after their appearance. Spots are seldom seen within less than five or beyond thirty-five degrees latitude each side of the Sun's equator.

There is no cause yet assigned for the periodicity of the Sun's spots. It may be found to be only a succession of natural reactions within itself, where one extreme follows another, the same as often seen upon the Earth, or it may be found to be a tidal effect, caused by the rotation on its axis when affected by the attraction of some of the planets.

The Sun is also revolving around a common center, which is always within itself. It is also said to be traveling toward the constellation Hercules at a rate of more than a hundred and fifty million miles a year, but the location of the center of the force or power which propels it, aside from Deity, has not yet been determined.

Besides attracting new forces, or influences, as the Sun and planets are traveling along through the heavens, this force, which is continually being thrown off from the Sun and planets at and near their equators, is evidently nearly all attracted back to them soon again, and as rapidly and nearly as great in quantity as that which was thrown off, but towards a much larger surface which is at each side of the equatorial planes. Some of this force is returning contiguous, some adjacent and some quite remote from the stream which is being thrown off. Much of it which is returning contiguous to the stream is drawn into it again by eddies which are continually forming, and there carried away again before getting near the body which attracts it. The disturbances of the magnetic needle indicate currents which correspond to the action of the Sun's force at its surface, in passing from toward the poles of the Earth to the equator, but never from the equator toward the poles. A possible corresponding effect may be seen in the Aurora Borealis, or Northern Lights, as well as the Zodiacal Light, as seen in the tropics, the latter of which may be due to the action of the Sun's instead of the Earth's force. This action of the Sun and planets gives us a system which admits of the forces or influences being used over and over again, and which clears the space of planetary matter, as it becomes secured to the planets and Sun by their attraction whenever passing near enough to each other, thus ever maintaining a firmament in the heavens and limiting the calculations of the time of worldly or planetary existence to the Infinite Ruler of the Universe.

### MERCURY.

Mercury, as shown in Figures II and IV, is the nearest known planetary body to the Sun. Its diameter is about three thousand miles, and it travels once around the Sun, from west to east, in about eighty-eight days. It rotates on its axis, from west to east, once in twenty-four hours and about five minutes. It has an orbital velocity of about seventeen hundred miles per minute, and travels at an average distance from the sun of about thirty-five million five hundred thousand miles, and in an orbit which is more elliptical than any of the orbits of the principal planets, its least distance being about twenty-eight million one hundred and fifty thousand miles, while its greatest distance from the Sun is about forty-three million miles. Its orbit is inclined about seven degrees to the ecliptic, being more than twice the amount of the inclination of the orbits of any of the rest of the principal planets.

When it is in perihelion, or its nearest approach to the Sun, it is much nearer the Sun than it is natural for it to remain, for the repelling force of the Sun at Mercury's

FIG. IV.

SECTIONAL SIDE VIEW OF THE SUN WITH JUPITER,
SATURN, URANUS, NEPTUNE ETC.

perihelion distance considerably exceeds the attraction of the Sun for the planet, and so it is gradually borne farther away from the Sun while moving along in its orbit; while, at the same time, the momentum which it received when it was approaching, and when in, and near its perihelion, now aids the Sun's force in carrying it to its aphelion, or to the farthest part of its orbit from the Sun, at which distance it is beyond its mean place and where the attractive force of the Sun for the planet exceeds the repelling force of the Sun. After the planet leaves its perihelion it gradually loses its speed and momentum until it arrives at its aphelion. When traveling from aphelion to perihelion it is gradually drawn a little nearer to the Sun by the excess of the Sun's attraction over its repelling force, thus gaining speed and momentum in proportion until it arrives at its perihelion again. The momentum that it has gained or acquired since leaving its aphelion carries it nearer to the Sun again than its mean place, and so it is thrown off as before, and always in an elliptical orbit.

## VENUS.

Next to Mercury, we have the planet Venus, as shown in Figures II and IV. Its diameter is about seven thousand seven hundred and fifty miles, and it travels once around the Sun in about two hundred and twenty-four and three-fourths days, at an average distance of about sixty-seven millions of miles from the Sun, and in an orbit which is nearly circular. Its ellipticity is said to be only about nine hundred thousand miles. Venus travels with an orbital velocity of but little more than thirteen hundred miles per minute, and with its poles greatly inclined to the ecliptic, while its orbit is inclined to the ecliptic only three degrees and about twenty-three and a half minutes. It seems that it is to the throwing-off force of the Sun that we should look for the cause of its orbit being so nearly circular. If the Sun's force at its perihelion exceeds the attraction of the Sun for the planet but little, it is evident that the Sun would have but a short distance to force it away before the attraction and throwing-off force of the Sun would be equal, and it could only go the distance beyond its mean distance that the excess of the Sun's force over the Sun's attraction, when at perihelion, would send it, and the momentum gained or acquired on its return from its aphelion would be so little that it could get but little closer to the Sun at its perihelion than its average distance in its orbit. Thus we see that this law, as in the case of Mercury, would cause the small ellipticity of the orbit of Venus.

If the Sun should remain in one position and Venus should travel around it in the direct line of the Sun's force, we might expect that an equal force of the Sun would keep Venus at an equal distance in its entire circuit. But, as the Sun is traveling through the heavens at an immense velocity and taking all of the planets along with it, it would be in harmony with this law to presume that when Venus passes between the Sun and the point toward which the Sun is traveling, it would be a little nearer the Sun at that point and a little farther from the Sun when on the opposite side, for, in the first instance, the Sun would be in the act of approaching Venus, and in the second it would be in the act of leaving it, which might make the planet's orbit perceptibly elliptical. The question often arises whether or not we should take into consideration the movements and positions of the planets, in determining their specific gravity, etc.; for we see the planet Mercury plunging into the Sun's force and then being thrown correspondingly away, while Venus is gently drawn to a position where the force is at all times nearly equally applied. This seems to be plainly shown by the inclination of the orbit of Venus to the ecliptic, which is only three degrees and about twenty-three and a half minutes. As before stated, the greatest inclination of the plane of the Sun's equator is supposed to be a little more than seven degrees to the ecliptic, which is best seen about the twenty-first of March, at which time the Earth is the farthest below, or south of, the Sun's equatorial plane, and about the twenty-first of September, when the Earth has reached the point which is farthest north of the Sun's equatorial plane, or force line. The center of the Sun's force is then passing about seven degrees below, or south of, the Earth.

When the ascending node of Venus occurs in that part of its orbit which corresponds to the Earth, when nearing the middle of December, it crosses the Sun's force in nearly the same relative position in which the Earth does, and when it travels around in its orbit until it gets to a point which nearly corresponds to our Vernal equinox, it is about three and a half degrees north of, or above, the plane of the ecliptic, and south of, or below the Sun's equatorial plane, which would place Venus within about three and a half degrees of the center of the Sun's force. That is the greatest distance away from it that the planet Venus ever gets. The same conditions occur on the opposite side of the orbit in the relative position to our Autumnal equinox, except that Venus is about three and a half degrees above, or north of, the Sun's equatorial plane, instead of below it.

An extract from the *Mining and Scientific Press* of November 13, 1880, adds more proof of the active principle of this force, as will be seen by the perusal of the following:

"Mr. R. G. Jenkins, F. R. A. S., has endeavored to show
" a very remarkable effect of the planet Venus upon
" the Earth. The present British Astronomer Royal
" proved, many years ago, that the disturbing effect of
" this planet was so great that the Earth was mate-
" rially pulled from its orbit. Mr. Jenkins shows that
" it is to this action that we must look for an explanation
" of the cold waves which occur on an average every eight
" years, as in 1829, 1837, 1845, 1855, 1863, 1871, 1879, and
" that for the next fifty years the temperature will be below
" the average. He states that a heat wave has been ob-
" served to pass over the Earth every twelve years, nearly
" cotemporary with the arrival of the planet Jupiter at its
" perihelion, such a wave being now close at hand."     •

By reference to Figures II and IV it will be seen that when the planet Venus passes between the Earth and the Sun that considerable of the force of the Sun must be intercepted that would have acted upon the Earth. When the Sun's repelling force, as applied to the Earth, is partially diminished by the interception of the planet Venus, the Earth falls toward the Sun in proportion, and may also be attracted a little by Venus at the same time, but as soon as Venus passes along so as to leave the Earth exposed to the full force of the Sun, the Earth is thrown back again to its natural orbital distance. If an increased amount of the Sun's rays will supply an increased amount of heat, a cold wave would be the natural result when such rays were intercepted and diminished. It is the general belief that the nearer the Earth is to the Sun, the greater will be the heat furnished by the Sun. But, in this instance, when the Earth is drawn materially nearer the Sun, the heat of the Sun is sensibly diminished.

The heat wave to which the eminent astronomer refers, in connection with the planet Jupiter, is only another natural result that might be expected to follow those conditions, for the force that is required to repel the planet Jupiter and roll it on in its orbit must be very great, and as this force extends in straight lines from the Sun to the objects which it repels, no doubt this force is more compressed in the direction of any resistance than where it has free exit from the Sun, and while the Earth is passing through a region of this compressed force an increased temperature on the Earth's surface would be the more natural result.

### EARTH.

The Earth is also shown, in Figures II and IV, as the third planet from the Sun. It makes one orbital revolution in about three hundred and sixty-five and a fourth days, at an average distance from the Sun of about ninety-three million miles, in an orbit somewhat elliptical, and to which the Sun's equatorial plane is inclined about seven degrees, as before stated.. Its axis is inclined to its orbit, or ecliptic, about twenty-three and a half degrees, and it rotates on its axis once in twenty-three hours fifty-six minutes and about four seconds. By comparing the conditions and motions of the Earth with those of Mercury and Venus, it is obvious that it must also be controlled by the same force and in about the same manner, for when, in that part of its orbit farthest from the Sun (toward the last of June), it appears to be in the direct line of the Sun's force, or equator, as the spots on the Sun appear to travel straight across the Sun's surface, and when it reaches its autumnal equinox it gets so far to the north of the Sun's equatorial plane that the spots in passing across the Sun appear to descend and rise to an extent that would indicate that the Earth is about the beforementioned seven degrees to the north of the Sun's equatorial force line. This allows the Earth to be drawn a little nearer to the Sun and to gain a little in velocity, which continues until it arrives at its perihelion and in the direct line of the Sun's equatorial force again, as indicated by the spots appearing to pass straight across the Sun's surface. In this part of its orbit it is nearer to the Sun than its average distance. The Sun's force there exceeds the Sun's attraction for the Earth, and, as it passes along in its orbit, it loses speed and momentum, and is borne gradually a little farther away by the excess of the Sun's force, as in the case of the planet Mercury. When it arrives at its Vernal equinox the spot lines which cross the Sun appear bowed a little, showing that the Sun's greatest equatorial force was passing a little above or north of the Earth, at that point of its orbit. As it passes on from here it continues to lose its orbital velocity and momentum until it reaches its aphelion, thus completing one orbital revolution.

## AXIAL INCLINATIONS.

The inclination of the poles of the planets to their orbits may be due to one polar hemisphere being heavier than the other and, consequently, more affected by the attraction of the Sun, when in the more distant part of their orbits or where the Sun's attractive exceeds its repelling force for the planet. As the Earth attracts the heavier side of the Moon toward itself, so the Sun, according to the same law, may attract the heavier part of the planets toward itself, and the inclination of the poles of the planets, to their orbits, may indicate the degree to which they are unbalanced in form or attractive properties, Jupiter showing a nearly balanced and Venus a greatly unbalanced condition. We know that the northern hemisphere of the Earth is heavier than the southern, in proportion as the amount of the earthy matter of the northern exceeds that of the southern hemisphere in volume. It may be shown that it is the unequal degree of the Sun's repelling and attractive influences, as acting upon the Earth in connection with the Moon's attraction while the Earth is in different parts of its orbit, that causes a gyratory movement of the Earth's North Pole and, as a change of the direction of the Earth's North Pole implies a change in the plane of the Earth's rotation, also the direction of its throwing-off force, a shifting of the Moon's nodes on the ecliptic might follow, as a natural result, when the Moon is seeking its natural position in relation to the Earth's repelling force.

### THE MOON.

The Earth's satellite, or Moon, is about two thousand one hundred and sixty miles in diameter, and at a mean distance of about two hundred and thirty-eight thousand miles from the Earth. Its motions, in its orbit, are probably as complex as those of any of the primary or secondary planets of our solar system. It makes one revolution around the Earth, with respect to the Sun, in about twenty-nine and a half days, and also one revolution on its axis in the same time, and a little more than twelve and a quarter revolutions, around the Earth, in one year. Its orbit is quite elliptical, the distance of the Moon varying from about two hundred and twenty-six thousand to two hundred and fifty-one thousand miles from the Earth. It seldom makes two revolutions around the Earth in the same time in the same year, the time varying from a few minutes to several hours in every year, until after about eighteen years, when we expect the same conditions to occur again. Nor are any two quarter moons of the same duration. Sometimes the Moon passes from quarter to quarter in about six and one-half days, while at other times it requires as much as eight and one-quarter days, and oftentimes in the same year.

In the accompanying diagram is shown, as near as proportions will conveniently allow, the path of the Moon

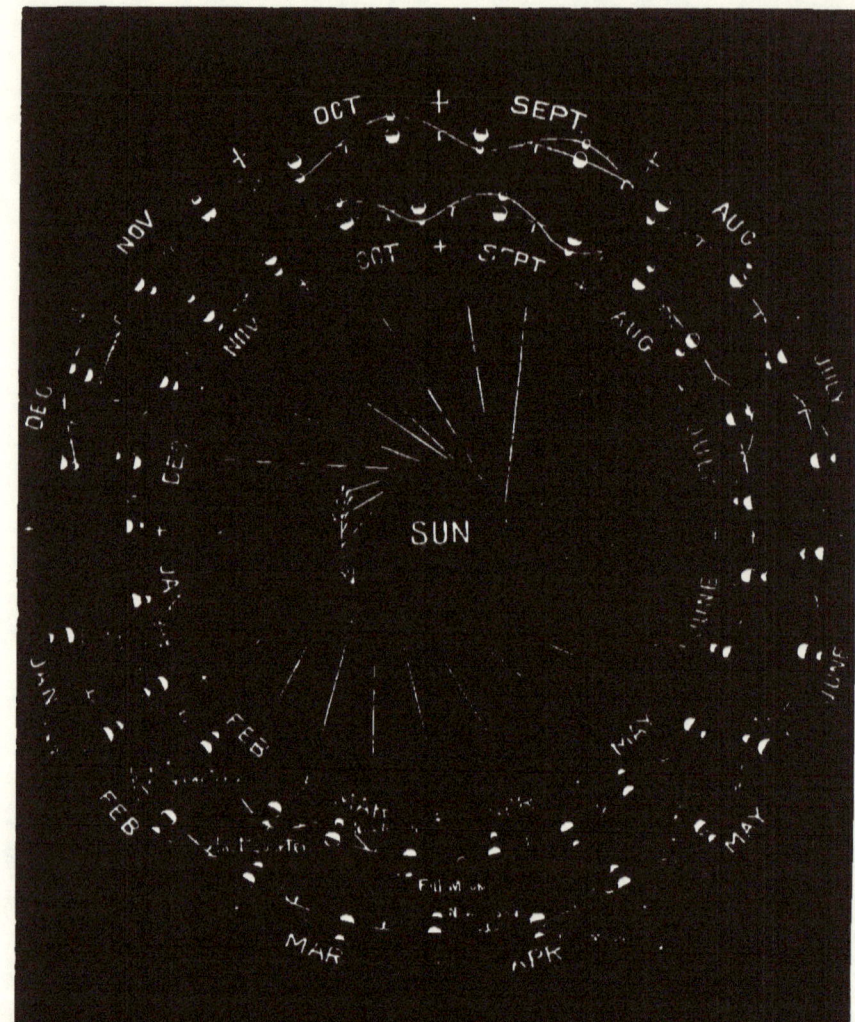

THE MOON'S PATH SHOWN ALONG THE
EARTH'S ORBIT DURING THE YEARS
1888 AND 1889.

crossing and recrossing the orbit of the Earth during the years 1888 and 1889; also the positions of the Moon in its new and full moon, and markings on the Earth's orbit showing the position of the Earth when the Moon is crossing the Earth's orbit just in the rear of the Earth at the time of the first quarter, and the position of the Earth when the Moon crosses the Earth's orbit just in advance or ahead of the Earth at the time of its last quarter. The path of the Moon along the Earth's orbit is slightly scolloped, it deviating from a straight line, or from the Earth's orbit, only about one mile in twenty, or two hundred and thirty-eight thousand miles in the distance that the Earth travels in its orbit during the time from first quarter to full moon, with the two hundred and thirty-eight thousand miles added, which the Moon must gain on the Earth during that time. The distance is sometimes more or less to the mile than stated. If the Moon when nearest the Earth, or when in perigree, is about the time of being new or of appearing full, the deviation is less, and when the Moon is in apogee in corresponding time, its deviation is more. The points of perigee and apogee are not fixed in its orbit with relation to the Sun or Earth, but are steadily advancing, making an entire revolution on the Moon's orbit in eight years and about two hundred and eighteen days. At the time of new moon the Moon passes between the Earth and the Sun. The Moon is in or near the Earth's orbit when it appears about half full, the quarter after each new moon being the first quarter and the quarter preceding the new moon being the last quarter, both of which occur while the Moon is on the inside of the Earth's orbit, or between the Earth's orbit and the Sun. When the Moon is on the opposite side of the Earth from the Sun, we see nearly the entire surface that the Sun shines upon. It is then full moon, and the quarter before the full moon is the second quarter, and the quarter after the full moon is the third quarter, both occurring while the Moon is outside of the Earth's orbit. During the year of 1888 the Moon was more than four days more time on the inside than on the outside of the Earth's orbit, and during the year 1889 it will be about nine and a half days more time inside than outside of it. And, although during the twelve complete moons from new to new, in 1889, the Moon is more than five and a fourth days more time inside than outside of the Earth's orbit, yet it travels along the Earth's orbit nearly a million more miles on the outside than on the inside of it, owing to the fact that it must gain twice the distance that it is from the Earth during the second and third quarter of each Moon, or each revolution around the Earth. When the Moon is new, or directly between the Earth and the Sun, its form and surface markings are faintly shown by the sunlight that is reflected from the Earth back to it. At that time the Earth would appear round or full, if observed from the Moon, and when the Moon is full, the Earth being between the Moon and the Sun, would appear as the new moon, except that the Earth would appear much larger and the surface much more dim, owing to the small amount of light that the Moon would reflect back to the Earth.

When the Moon is at first quarter it is in the act of crossing the Earth's orbit going from the inside to the outside of it, and when in that position we see only one-half of that half of the Moon that the Sun shines upon, causing it to appear half full, or at first quarter. If the Earth could be seen from the Moon at that time, it would appear to be half full, also, and as the quarter after the full moon, or last quarter. Although the Earth is traveling in its orbit at the rate of about one thousand one hundred and ten miles per minute, the Moon must gain its average distance after the first quarter, or about two hundred and thirty-eight thousand miles on the Earth, before it can appear as full moon; and from there it must continue to gain, until it crosses from the outside to the inside of the Earth's orbit, which is about two hundred and thirty-eight thousand miles, and directly in front of, or in advance of the Earth, at which time we see only one-half of that part of the surface of the Moon that the Sun shines upon, and which completes the third quarter. From here it travels ahead of the Earth, gradually getting to one side and also slackening its speed, until the Earth has caught up with it in relation to the Sun, when it appears directly between the Earth and the Sun, or at new moon, which completes its fourth or last quarter. From the position of new moon it is gradually drawn back to the Earth's orbit, or at first quarter, and with an accelerating speed corresponding to the decreasing speed of the fourth quarter. It should be observed here that the Moon has to lessen its speed only a very little during the fourth and first quarters in order that the Earth may pass it, for the Earth during these two quarters travels nearly twenty-two and three-quarter million miles in its orbit, and only has to gain twice the distance of the Moon from the Earth, or less than four hundred and eighty thousand miles. The speed of the Earth in its orbit may be pretty well realized by reference to the position of the Moon at the last of the third or beginning of the fourth quarter, when it is one-half full; as it is then on, or directly in, the line in which the Earth is traveling in its orbit, and at a point to which it takes the Earth only about three and a half hours to arrive. As the Moon's orbit is not parallel with the Earth's, but inclined about five degrees to it, the Moon crosses and recrosses this plane or ecliptic twice in each revolution around the Earth. The point in its orbit where it descends from above, or north of the ecliptic, to below, or south of it, is called the descending node. And the point in its orbit which intercepts the ecliptic when it is passing from below, or south of the ecliptic, to the region above it is called the ascending node. These points of intersections are not fixed in the Moon's orbit, but occur a little farther to the West at each succeeding revolution of the Moon around the Earth, and are called the retrograde movement of the nodes. They make an entire revolution on the orbit of the Moon in eighteen years and about two hundred and nineteen days. The Moon presents nearly the same surface to the Earth all the time, which is said to be owing to the fact that the central point of attraction upon which the Earth is acting in the Moon is a little nearer to the side of the Moon which is turned toward the Earth. The side of the Moon which is constantly presented to the Earth is thought to be considerably the heavier, the reason for which is yet speculative.

Whether the Moon ever revolved on its axis more than once in each revolution around the Earth is not known. If it ever did, its process of formation or volcanic action during its cooling and hardening process may have changed its form a little, thereby shifting the center of gravity until it assumed its more recent or present motions. The revolutions of the Moon on its axis are claimed by some to be absolutely uniform. Yet we know that the Moon travels faster in some parts of its orbit than in others. When near perigee it often travels from quarter to quarter in a little more than six and a half days; when near apogee it often takes nearly eight and one-quarter days to go from quarter to quarter, or to make one-fourth of one revolution on its axis. Sometimes an extremely slow quarter precedes a fast one, and sometimes a fast one precedes a slow one. If it were revolving on its axis at the slow rate, or once in about thirty-three days, and quickly got into conditions where it would need to revolve on its axis once in about twenty-six days, the attraction of the Earth on the Moon might not be sufficient to change the velocity of its axial rotation so suddenly, and it would seem to pass along or ahead of its mean place and allow us to see a little of the farther half of it. Or, if its axial rotation occurred once in about twenty-six days, its momentum might revolve it a little beyond its mean position, at times when its axial velocity would be reduced to one revolution in about thirty-three days. In each case we would be enabled to see a little more of the Moon, first on one side and then on the other, than when in its mean position. If the orbital revolution of the Moon should change from its present time to either twenty-six or thirty-three days, the Earth's attraction would evidently change its axial rotation accordingly.

The axis of rotation of the Moon is inclined about one and one-half degrees to its orbit, which, taken in connection with the inclination of its orbit to the ecliptic, makes it appear at times to tip forward and backward about thirteen degrees, which enables us to look a little over its North Pole, and when in opposite positions a little over its South

Pole. If the Moon is observed first from one side of the Earth and then from the other at proper times, a little more than one-half of the Moon can be seen, on account of the Earth being so much larger than the Moon we see a little over the edge of it. But, if viewing the Moon from any one point, not taking into account its oscillating movements, we never see quite one-half of it, owing to the lines of vision starting from one point and not being able to reach over to an imaginary diameter of the Moon, as placed or located at right-angles with the line of observation. These several apparently oscillating movements of the Moon are called its librations.

When we take into consideration the immense speed at which the Earth is traveling around the Sun, the several motions of the Moon, together with the rapidity with which it passes from one position to another, and the influence of the attraction of the Earth and Sun, also the throwing-off force of the Sun, etc., it will be seen how difficult it is to account for its various and multiple positions without the aid of apparatus to illustrate them.

The conduct of the Moon in going once around in its orbit, when carefully noted in its entire circuit, seems to exhibit more positive signs of the throwing-off force of the Earth than any other example which is given us among the planets. If, starting with the full moon, we find it outside the Earth's orbit, and in a position relative to the force of the Earth (as shown in diagram), that would naturally cause it to gain on the Earth, as it appears to be continually driven along and ahead by the Earth's force, until it is carried around in front of, or in advance of the Earth, where it crosses the Earth's orbit at the beginning of the fourth quarter, and is still driven ahead and a little to one side, where its position is such that the tangent force lines from the Earth would not have the effect of sending it on still ahead of the Earth, but would rather appear to push it to one side and to partially or gradually hold it back and to bear it away until the Earth passed it sufficiently to allow it to be drawn toward the Earth by attraction, and at the same time press it behind, or in the rear of the Earth, by the Earth's throwing-off force. From there it soon gets into harmony with the direct force rays of the Earth, and is enabled to gain upon and to pass the Earth again at full moon. Owing to the different degrees of the Earth's force that the Moon is traveling in, the points of apogee and perigee vary considerably in the Moon's orbit. Sometimes the Moon passes from one point to the other in about twelve days, while in the following or preceding year and in the corresponding Moon, it would take about sixteen days. On the average the Moon is thrown a little further round at each or every revolution, causing the points of apogee and perigee to advance in the Moon's orbit.

To this same force may be assigned the principal cause for the inclination and change of the Moon's orbit as shown by its nodes, which are gradually falling back probably because it is not thrown quite as far to the south as it is to the north of the Earth's force, and it gets back to the plane of the ecliptic a little sooner each time, which action would cause the retrograde movement. The intervals between the nodes are much more uniform than between the apogee and perigee. According to M. Ligner, the Austrian meteorologist, the Moon has an influence on a magnetic needle, which appears to be greatest when the Moon is in the plane of the Earth's equator, and also greater when the Moon is south than when it is north of the ecliptic.

## MARS.

Mars is also shown, in Figures II and IV, as the fourth planet from the Sun, revolving on its axis and around the Sun from west to east. Its diameter is about four thousand two hundred miles. Its orbit is considerably more elliptical than the Earth's. Its mean distance from the Sun is about one hundred and forty-one million miles. According to observation its axial rotation is twenty-four hours thirty-seven minutes and about twenty-two seconds. About six hundred and eighty-seven days are required for one orbital revolution. Its axial equatorial velocity is only about five hundred and thirty-six miles per hour, while our Earth's velocity is a little more than one thousand. Therefore it appears that the equatorial force of Mars must be much less than that of our Earth. Mars has two small satellites traveling from west to east, and, according to the more recent authors, their orbits are about circular and lie nearly in line with the planet's equator. Phobos, or the inner satellite, is about eleven miles in diameter, and travels around its primary at an average distance of about five thousand eight hundred miles in seven hours and nearly forty minutes, thus making a little more than three revolutions around the planet while the planet rotates once on its axis. This would cause the satellite to rise in the west and set in the east, if observed from Mars, and this would make it appear to meet the stars, the Sun and the other satellite. Its orbital velocity is about eighty miles per minute, not taking into account the distance that it travels along the orbit of its primary.

The outer satellite, Deimos, is about eight miles in diameter, and about fourteen thousand miles from the planet. It makes one orbital revolution in thirty hours and about eighteen minutes. Its orbital velocity is about forty-seven and a half miles per minute, while our own Moon only appears to be traveling at the rate of about thirty-five miles per minute. If the paths of Phobos and Deimos were shown in diagram along the orbit of Mars, they would be seen crossing and recrossing it in a similar manner to that of our Earth's satellite, except that the intervals of crossing would appear nearer together and a little more scolloped. As Mars revolves on its axis once in about twenty-four hours, any given meridian on its surface would gain on Deimos so slowly, that it would appear to rise very slowly in the east, and after a little more than five days would appear to set slowly in the west.

Figures II and IV also represent Mars as being subject to the same influence and controlled in like manner by the same forces as the planets between it and the Sun. The throwing-off force is well illustrated by Mars and its satellites. As Mars has an equatorial axial velocity of only about five hundred and forty miles per hour, the force that it develops by its rotation must be proportionately less than those which have a greater equatorial velocity. This is shown by the near approach of the satellites to the planet, it appearing scarcely able to keep the inner satellite from joining it. Although this force is apparently small, yet it seems to be clearly shown to be in operation when the satellite Phobos is crossing the planet's orbit only five thousand eight hundred miles in front, or in the advance of, its primary, for it must be speedily carried or urged along and to one side, for it takes Mars only about six minutes to travel in its orbit the distance that Phobos is from it. After crossing the orbit in front of the planet the satellite gradually slackens its speed as Mars passes along, and then recrosses the orbit in the rear of the planet, then gains on and overtakes the planet in relation to the Sun at a point corresponding to our full moon, and from there it continues to gain until it arrives at the orbit in front of the planet again, thus making an entire revolution in seven hours and nearly forty minutes. The outer satellite is being operated upon in the same manner, although its movements are not so rapid.

If we should presume that there is a gaseous, electric, or other influences surrounding the planets Mercury, Venus, the Earth and Mars, and which, if extended far enough from their surfaces so that when the Sun's force would act upon them at a distance, that would roll them, together with the gaseous envelope, along in their orbits instead of swinging them, they would require gaseous depths from the planets about as follows, in order that they should roll around the number of times that they are said to rotate on their axes, while going once around the Sun: Mercury, about 409,000 miles; Venus, about 270,000 miles; the Earth, about 250,000 miles, and Mars, about 209,000 miles. These decreasing depths that would appear to be required with each planet, correspond to the decreasing orbital velocity as regards the distance of each respective planet from the Sun.

### ASTEROIDS.

All of the Asteroids so far discovered, and which now number about two hundred and seventy, revolve in orbits between the orbits of the planets Mars and Jupiter, and appear to obey the same general law as regards their movements from west to east around the Sun, as well as in the ellipticity of their orbits. Of their axial rotations nothing is known. Only a very few of them are ever seen without the aid of instruments, and not much interest is attached to their discovery, or even to their existence, except to the astronomical student. Their names, times of orbital revolution, many of their diameters, inclination of their orbits to the ecliptic, etc., are tabled in most of the text-books so completely that a reference to their movements as being in harmony with the general law or force, as previously described and illustrated, is deemed sufficient.

### JUPITER.

Jupiter is shown in Figure IV as the fifth planet from the Sun, and controlled by the same force and influence as the planetary bodies previously described. It is the largest planet of the solar system. Its equatorial diameter is claimed by some authors to be about eighty-five thousand miles, and by some as great as ninety thousand miles. The polar diameter is estimated to be about fifty-five hundred miles less. After many careful observations, authors generally agree that it rotates on its axis once in nine hours and about fifty-five minutes. It travels around the Sun in about four thousand three hundred and thirty-two and one-half days, at a mean distance of about four hundred and eighty-three million miles, in an orbit considerably elliptical and inclined one degree and about eighteen minutes to the ecliptic. Its axis is inclined only about three degrees to its orbit; therefore there would be scarcely any change in its seasons. The excess of the equatorial over the polar diameter is, without doubt, caused by its rapid rotation on its axis, together with the tidal action and effect produced by its moons. Belts are seen to extend across the planet parallel to its equator, and which are constantly changing, sometimes increasing or decreasing in number and size, and sometimes appearing to change color. Spots frequently appear on its surface, and by them the time of its axial rotation has been determined. The spots near the equator have the same peculiarity of traveling faster than those farther from it, as they do on the Sun, which is probably owing to the tidal effect of the moons, as they are all nearly in line with the plane of the planet's equator. If the influence of the moons of Jupiter produce an effect on its surface similar to our own satellite on the Earth's surface, the movable substances near the equator of Jupiter would be correspondingly changed in their positions, while the spots farther from it might be moved along a little in proportion to the tidal effect, which would imply that the spots which are traveling the slowest are traveling somewhat faster than the body of the planet really revolves. The fact of there being a difference in the speed of the spots on the planet's surface is evident that the planet itself is either traveling faster than the fastest spots, and causing them to follow, according to the centrifugal effect of the planet, or slower than the slowest, and by its greater throwing-off force near the equator, in connection with its moons, is urging the spots along at that part of its surface more rapidly. It seems very much at variance with the law of mechanics that there should be a medium speed between the times of the rotation of the spots at which the motive power is moving.

As all of the planets lie nearly in the Sun's equatorial plane, or in nearly the same relation to the Sun as the moons of Jupiter do to it, so it would appear that the same cause or principle which would advance a spot on one could also be claimed to advance it on the other.

The Sun's force seems to be applied differently to the planets which revolve in their orbits outside of, or beyond, the asteroids, while it seems to be controlling the four inner planets by acting upon the gaseous fluid or influence which surrounds them at various distances from them. The force appears to be applied almost, if not quite directly, to the surfaces of the four which are beyond the asteroids. For, if we presume Jupiter to be rolled around in its orbit, with no allowance to be made for its being in an elastic medium, we would have the number of rotations that it would make in one orbital revolution by dividing its orbital distance by the circumference of the planet. Then divide the time of one orbital revolution by the number of rotations obtained, and we shall have the time for each rotation on its axis. The result thus obtained varies from nine hours and about nineteen minutes to nine hours and about forty-two minutes, according to the diameters and distances given by different authors. Diameters given by one author and distances given by another could be selected, which would give results that would be quite near to the time given by those who have made careful observations. Jupiter's orbital velocity is about five hundred miles per minute, and its equatorial axial velocity is about for hundred and sixty miles per minute.

## MOONS OF JUPITER.

Jupiter has four moons traveling around in orbits from west to east, and at distances from the primaries about as follows, respectively: Two hundred and sixty-seven thousand miles; 425,000 miles; 687,000 miles, and 1,193,000 miles, and whose orbital revolutions occur in one day eighteen hours and about twenty minutes; three days thirteen hours; seven days three hours and forty minutes, and from sixteen to eighteen days for the fourth, or farthest one. Their orbital velocities also decrease as their distances increase, from about seven hundred miles per minute for the nearest, to about three hundred and ten miles per minute for the farthest. All of the moons cross and recross the orbits of Jupiter in the same manner as those of Mars and our own, except that if illustrated their paths would appear much more scolloped and the crossings much nearer in proportion, while the inner one must very nearly make a loop in its orbit when passing between Jupiter and the Sun.

The positions, etc., of all of the moons of Jupiter seem to confirm the theory in regard to the manner and direction of the repelling forces, as heretofore designated. The orbits of the moons of Jupiter correspond very nearly to the orbit of Venus in relation to the Sun, in that they are all nearly in line with the equatorial plane, and all nearly circular.

The inclination of a planet's orbit to the Sun's equatorial plane, or force line, does not always determine the ellipticity of the planet's orbit, because the planet may be traveling so far away from the Sun that a less angle would carry it farther to one side of the Sun's force line than planets which are nearer with a greater orbital inclination to the Sun's force.

## SATURN.

Saturn is also shown in Figure IV. It is the sixth principal planet in distance from the Sun, and the farthest one of our solar system which was known to the ancients. The author has never seen any evidence of any knowledge by them of its ringed appearance. Saturn is the next largest in size of all the planets belonging to the solar system, it being from about seventy thousand to seventy-four thousand miles in diameter, according to different authors. Its mean distance from the Sun is variously stated at from eight hundred and seventy-two million miles to nearly eight hundred and eighty-eight million miles. Its poles are inclined about twenty-eight degrees to the ecliptic, and late authors give ten hours and about fourteen minutes as its axial rotation, and about twenty-nine and a half years as its orbital revolution. Its orbit is considerably elliptical, and but little inclined to the ecliptic. Changeable bands or belts appear to surround the planet, parallel to its equator, similar to those of Jupiter, and spots have also occasionally been seen on its surface, by which its time of rotation has been determined. Although Saturn does not revolve as fast as Jupiter, yet its equatorial diameter is supposed to be greater in proportion; but it corresponds to the almost continual tidal effect which its more numerous and oftener-appearing moons would produce, although there is not much difference between their combined bulk and those of Jupiter.

### MOONS OF SATURN.

Saturn has eight moons, revolving at distances varying from nearly one hundred and twenty-one thousand miles to nearly two million five hundred thousand miles from the primary, and with a decreasing velocity respectively from about five hundred and sixty miles to about one hundred and twenty-seven miles per minute, not considering the orbital velocity of Saturn. Nearly all of their orbits lie close to the line of the equator of Saturn, which circumstance would indicate that their orbits are nearly circular, or similar to the orbits of the moons of Jupiter and the orbit of the planet Venus. If a wheel were rolled along the side of a vertical plane surface with points fixed at various distances between the center and the rim of the wheel for marking on this surface, the lines produced would somewhat represent the paths of the satellites along the orbits of their primaries. One placed at a point about three-fourths the distance from the rim to the center of the wheel, would nearly represent the path of our own Moon along the orbit of the Earth, while those nearer the rim would better represent the paths of the satellites of Jupiter and Saturn. The straight line caused by the axis of the wheel would represent the orbit of the planets. The inner satellite of Saturn nearly, if not quite, makes a small loop in its path while passing between its primary and the Sun at each revolution around the planet. If the orbital distance of Saturn be divided by the planet's circumference to obtain the number of rotations in going once around the Sun, and the time of one orbital revolution be divided by the number of rotations thus obtained, we shall have from ten hours and about thirteen minutes to ten hours and about thirty-eight minutes as the time for Saturn to revolve once on its axis, according to diameters, distances and times given by several of the latest authors, extreme results either way being omitted.

### RINGS OF SATURN.

Saturn is surrounded with several rings of a material which seems to reflect the light of the Sun as it shines upon them. All of the rings lie in the equatorial plane of the planet. The outer diameter of the outer ring is about one hundred and sixty-six thousand miles. The outer rings appear very thin when the edge is presented to the Earth, while near the planet they seem to be a little thicker.

Some claim to have observed spots on the rings by which they presume the time of their rotation to be but little less than that of the planet. The manner in which the rings are suspended at different distances from the planet indicates that they are of different substances and of different degrees of specific gravity. This also shows the force of the planet acting through one substance and upon another beyond. The rings of Saturn furnish us the best ocular demonstration, in the solar system, of the throwing-off force of the revolving planetary bodies. Although the rest of the planets may not gather from space or throw off from themselves a substance that reflects or emits light, yet the action or force which produced the rings and still continues to hold the substances of which they are composed suspended in the position where the operation of the law placed them, must have been developed long before the rings were formed, or at the times when the planets commenced to rotate on their axes.

All substances or influences before being thrown off from a revolving body or sphere by centrifugal force seek that part near the equator, and which is farthest from its axis, and where the axial velocity is greatest, before being thrown off, and, of course, cannot be returned by attraction to the body at that part of its surface; but after the centrifugal force of the body has ceased to effect the substance or influence that is thrown off, then should it return to the planet by the planet's attraction, it must do so on one or the other side of this force which sent it away, and which is constantly kept up by the rotation of the revolving planet.

### URANUS.

Uranus is the seventh principal planet from the Sun, and is so far away that it is seldom seen with the naked eye. Its diameter is about thirty-three thousand miles. It travels from west to east around the Sun in an elliptical orbit which lies about in line with the Earth's, and at an average distance of about one billion seven hundred and eighty-five million miles from the Sun. Its orbital velocity is about two hundred and sixty miles per minute, and it makes one revolution around the Sun in about eighty-four years. The time of its axial rotation is unknown. Some of the later writers have placed it at about seven hours, and some from nine and a half to ten hours. If calculated by dividing its orbital distance by the planet's circumference, etc., in the same manner as Jupiter and Saturn, we shall have about seven hours as the time required for one axial rotation.

### NEPTUNE.

Neptune is the eighth and farthest planet from the Sun yet discovered. It is so far away that it is never seen with the naked eye. Its diameter is about thirty-seven thousand miles. It travels from west to east around the Sun, at a mean distance of about two billion eight hundred million miles, once in about one hundred and sixty-four years, with an orbital velocity of about two hundred miles per minute. Its orbit is elliptical, and inclined but little to the ecliptic. The author has seen no estimates of the time of its axial rotation. About nine and a half hours would be the result if calculated by its circumference and orbital distance, etc., as with Jupiter, Saturn and Uranus.

### MOONS OF URANUS AND NEPTUNE.

Uranus has four Moons and Neptune one. They are said to move in orbits greatly inclined to the ecliptic, and in a retrograde manner, or from east to west, around their primaries, motions which are contrary to all the rest of the planetary bodies of our solar system, the cause for which will probably be explained as science advances.

The points of intersection of the orbits of the planets with the plane of the Sun's equator do not occur in the same place at every revolution of the planet in its orbit. They either advance or fall back on the Sun's equatorial plane similar to the Earth's satellite in regard to its nodes or apsides. Some of the planets vary more than others, but the point of intersection with the Sun's equatorial plane is according to the extent that the Sun's force sends them away after leaving their perihelion.

MEAN DISTANCES.

The mean distances of the planets from the Sun are obtained by adding the perihelion and aphelion distances together and then assuming one-half of the amount thus obtained to be the average distance of the planet from the Sun. (See Mean Distance, Lockyer's Index.) The point thus obtained is one which corresponds to the central point of an ellipse, a position that the Sun never occupies in relation to any of the planets. It is always nearest the perihelion side of the orbit. The position thus occupied by the Sun in relation to the Earth is best seen at the time of the equinoxes. If an imaginary line were drawn across the Earth's orbit at those times and to those points, it would intercept the Sun, but would not be far enough from the Earth's perihelion to reach the central point of the long diameter of the Earth's orbit. A line from the perihelion to the aphelion points of a planet's orbit shows nearly the line of the long diameter of the orbit, but the line from equinox to equinox does not show the short diameter of the elliptical orbit, as the line which intercepts the equinoxes is always nearer the perihelion than the line of the short diameter. In mechanics, the sum of one quarter of the short diameter and one quarter of the long diameter of an ellipse added together, gives about one-half of the average diameter of an ellipse, and one-half of the sum of two unequal distances, which make a diameter, gives us only the radius of a true circle. The mean distances of the planets from the Sun, as given by various authors, is probably not far from being correct, for the orbits of the planets are probably nearer circular than they are generally stated to be. If we presume the orbit of Venus to be a true circle, with the Sun placed far enough to one side of the center of the orbit to indicate the ellipticity, as claimed by authors, or so that Venus is nine hundred thousand miles nearer the Sun at perihelion than at aphelion, one side of the Sun would then be within twenty thousand miles of the center of the orbit. It is obvious that if we presume the circle to then be flattened so that the short diameter will be nine hundred thousand miles less than the long diameter, the entire half of the orbit on the perihelion side of the Sun would be about the same distance from the Sun, conditions which are entirely unnatural, and which no astronomer admits.

The Earth is traveling around in its orbit, with the Sun not two million miles distant from the center of the Earth's orbit. If we should presume the orbit to be a true circle with the Sun at that distance from the center, it seems clear that the orbit would admit of but very little ellipticity, else the Earth could not increase its distance much from the Sun between the time of its perihelion and its arrival at its equinox. The following table is deduced from the table of distances, etc., of some of the latest calculations:

TABLE OF THE APPROXIMATE DIAMETERS OF THE SUN AND PRINCI-
PAL PLANETS, SOLAR DISTANCES, ETC.

| | MEAN DI-AMETER IN MILES. | MEAN SOLAR DISTANCE IN MILES. | ORBITAL REVO-LUTIONS. | ORBITAL VE-LOCITY PER MINUTE. | AXIAL ROTA-TION AT SURFACE. | EQUATORIAL AXIAL VEL. PER MIN. |
|---|---|---|---|---|---|---|
| Sun..... | 866,000 | ......... | ......... | ......... | 25d. 0h. 0m. 0s. | Ab't 76 miles |
| Mercury | 3,000 | 35,500,000 | 88 days. | 1,700 miles. | 0d. 24h. 5m. 0s. | " 7 " |
| Venus... | 7,750 | 67,000,000 | 224½ " | 1,300 " | 0d. 23h. 20m. 0s. | " 17½ " |
| Earth... | 7,918 | 93,000,000 | 365¼ ' | 1,110 " | 0d. 23h. 56m. 4s. | " 17 " |
| Mars.... | 4,200 | 141,000,000 | 687 | 960 " | 0d. 24h. 37m.+s. | " 9 " |
| Jupiter.. | 88,000 | 483,000,000 | 4,332½ | 500 " | 0d. 9h. 55m. 0s. | " 460 " |
| Saturn .. | 72,000 | 885,000,000 | 29½ yrs. | 360 " | 0d. 10h. 14m. 0s. | " 368 " |
| Uranus.. | 33,000 | 1,785,000,000 | 84 " | 260 ' | About 7 h. | " 242 " |
| Neptune. | 37,000 | 2,800,000,000 | 164 " | 200 " | About 9½ h. | " 204 " |

## COMETS.

The supposed orbits of some of the most noted comets are also shown in Figure IV. The application of the same law to the comets as to the planets is shown by the radiating force lines. It is well known to astronomers that the tails of comets are often not formed until sometime after the discovery of the comets, and also that the tails extend in a line from the comets on the opposite side of them from the Sun, also that the comets approach the Sun from the space at either side of the plane of the orbits of the planets, and not from that region of space near the Sun's equatorial plane or where the Sun's force is greatest, as shown in the sectional edge-view of a part of the solar system in Figure II. Figure IV also shows the comets approaching and receding from the Sun. If approaching from either side, they may be under the radiating influence of the Sun sufficiently to form the tail of the comet while at the same time the force of the Sun may not be strong enough at the side or in the region of the comet to throw it away, so they approach nearer and nearer until drawn sufficiently into the current to be thrown away again. In this manner, they make their varying and elliptical orbits as shown, the ellipticity of their orbits probably depending, to a great extent, on their composition and the extent to which they are drawn into the Sun's force.

The curvature of the tail of the comet is shown as being caused by the force of the Sun carrying particles of the cometary matter or substance off in the directions of the Sun's force lines, which after a time are disseminated, while others are continually making their appearance as the comet passes along in its orbit.

The great comets that came in 1881 and 1882 appeared so suddenly and unexpectedly that they were not seen by many until the maxima of their tails were nearly reached, thus showing that they entered the radiating force of the Sun in a short period of time, which could be done only from the side of this force where they would be affected but little until when drawn into the force they quickly appeared with luminous tails, caused by contact with the Sun's repelling force in the manner shown and as described. The tails of comets may not always be seen as formed by the Sun's force acting upon them in straight lines, for they are not seen in the direct and full force rays of the Sun, but more or less to one or the other side where the force is weaker and where it seems to be falling out of or leaving the main current in irregular lines.

Comets are sometimes known to approach and to travel around the Sun in an opposite direction from which the

planets are traveling in their orbits. Nearly all of the
comets having periods of return of about one hundred years
travel around the Sun in the same general direction as the
planets, or from west to east. The comets which do
appear more regularly and which have the same general
direction as the planets, might be presumed to have a
greater specific gravity than those which have longer
periods of return, or it may be that the direction or angle at
which they approach the Sun is so small that the momentum which the comets have gained by the attraction of the
Sun would not be sufficient to cause them to enter far
enough into the Sun's force, at the time of their supposed
regular visitations, to materially change the orbit of the
comets and their time of return. It also seems but natural that a comet which has been returning quite regularly
might, for some reason, approach the Sun at a greater angle
on some occasions than on others and by its momentum,
gained while returning, be drawn further into the Sun's
force than at other times and then be thrown correspondingly farther away, which would delay its next return to the
Sun. Some of the comets seem to keep in tolerably well-
defined and regular orbits, while some few which are
claimed to not return are said to have orbits as described
by the hyperbola. It seems in harmony with this law, if
the direction which the comet shall have already attained or
acquired should lead it into the Sun's force more and more
as it is receding from the Sun, that the comet would assume
a more and more direct line from the Sun, and would
continue in line with the force until the force became spent
in space, when the comet would cease to go farther, and from
that point might return to the Sun again from either side
of the stream of force which sent it away. It has been
clearly seen that some comets diminish in size as they
approach and recede from the Sun until they are lost from
view, and then on their return they seem to have increased
in size again, only to be again reduced as they approach and
depart from the Sun, conditions which indicate that the
substance which the comet attracted to itself while absent
is not so dense that the Sun's force cannot overcome the

attraction of the nucleus of the comet for the particles or substance which surround it and of which the body of the comet may be composed.  As these particles are carried off, they usually appear luminous and form the tail of the comet, as previously described.

As the planets have attracted to themselves different substances from the region of space through which they travel, so it would follow that a comet while absent would not necessarily attract or take up the substances of which it is composed every time alike, but would absorb such substances as it passed near enough to attract to itself, and when reappearing it might be seen with the same nucleus as before, but would show a somewhat different composition than it did on its preceding visit.  If a comet is small and lacks density it may be entirely worn away and so carried off into space by the Sun's force rays, under which circumstances its identity must forever disappear, while the substance of which it was composed might be absorbed in part by other comets, if not by planetary bodies.

The tails of the comets are sometimes formed at immense velocities, which, it seems, could not exceed the combined velocities of the comet going toward the Sun in one direction and of the particles or substances which are carried in the opposite direction by the Sun's force.  The speed with which the particles are carried away from the comet furnishes the best means of telling the speed of, and the distance from, the Sun's surface, at which the Sun's force is thrown off.

It would seem, from the easy manner in which the Sun appears to control the comets and the power that it has over them, that they could never come as a body in contact with any of the planets, because the Sun's force is so great in the region of the planets that a comet would be thrown back into the space beyond again, if it were to approach the Sun from that direction; and again, the force which is developed by the rotation of the principal planets on their axes and which keeps their satellites in their orbits, would be sufficient to repel a comet if approaching from near the planet's equatorial plane.  If approaching from either side of this force, the greatest attraction would be that of the Sun, to which it would be drawn as near as the Sun's throwing-off force would allow.

As the Sun is traveling through space, it might attract to itself a cometary body or substance from the deep, and one with which it never came in contact before, and then throw it off to the farthest limit of the Sun's force, there to be lost to us and never to return, which it seems would be the more likely result if thrown in the direction contrary to which the Sun is traveling in the heavens.

A comet might be drawn farther into the stream of the Sun's force than otherwise by the attraction of some of the planets when passing in the region of them, and if receding from the Sun would consequently be thrown farther away, making a more elliptical orbit, and also extending the time for the comet's return. These seem to be the circumstances attending the comet which appeared in 1779, the orbit of which indicated short and quite regular periods, when it was drawn by the attraction of Jupiter well into the Sun's force, and its orbit so changed that it has not since been seen.

When a comet is approaching the Sun its velocity is constantly increasing, until it is thrown off again by the Sun's force, when its velocity decreases, until the force is spent which sent it away, after which it usually returns to the Sun, and in the same manner as before. As the ball which is thrown in the air will return with the same velocity as that with which it ascended, so the planets and comets return to their perihelion with about the same velocity as that at which they were forced away.

The history of comets, as a whole, according to general observations, defines pretty clearly the direction of the force of the Sun, but gives us no idea of the eddies of, or the dividing and disseminating of, the Sun's force or current, at its extreme edge, other than the planetary and cometary action which gives evidence that the force is continually falling out of the stream at its sides, all the way along, from the center, or near the Sun, to its extreme edge, and then returns toward the Sun only to be thrown off again in a manner similar to the air, by a rotary blower.

### TIDES.

Among the many questions of more or less importance that arise while considering the motions of the Sun and planets by the foregoing theory and illustrations, that of the tides becomes one of much interest, for it is well known by those who have had opportunities to observe and investigate that the equatorial diameters of most, if not all, of the planets which revolve on their axes are considerably greater than their polar diameters, which is accounted for by no other theory than the result or effect of the centrifugal force that is developed by the velocity with which the planets rotate on their axes and the act of throwing off by this force the substance or material of which they are composed while in a fluid or movable condition farther from the center of the planets in the region of their equators than elsewhere on their surfaces. The tidal theory, which most authors have accepted and which is now generally taught in the later astronomical text-books, appears defective if the cause is examined in relation to the Moon, when in different parts of its orbit around the Earth. As commonly taught, the Moon draws the fluid portions, or the water, on the Earth on the side of the Earth toward the Moon, a little higher than the average height, thus making the higher tide, and at the same time drawing the Earth, not quite so much, but partially, away from the water on the opposite side from the Moon, thus causing the water to be drawn up to a ridge which forms the smaller wave on the opposite side of the Earth from the larger one. The objections to this theory are so numerous, and the theory is so much at variance with natural law, that only a few of the positions of the Moon in relation to the Earth will be referred to. It should be borne in mind that the Moon is traveling comparatively slowly and quite regularly around the Earth, going only about one twenty-ninth of its orbital distance, or between twelve and thirteen degrees of its entire circuit, in twenty-four hours, and which is only about thirty-five miles per minute, not taking into account the distance that it travels along the Earth's orbit in the same time. The attractive influence of the Moon on the Earth is in accordance with and in proportion to its distance from it. When in perigee the attraction is greatest, and least in apogee. The Moon arrives at these positions gradually at intervals of a little more than fourteen and one-half days.

If the theory in regard to the cause of the tides or the way they are produced, as popularly taught, is correct, then authors have apparently overlooked the fact that the Earth must be drawn a little for several days from its natural orbital position by the Moon at the time of the lunar tides, before and after the time of the new and full moon, and also must be drawn correspondingly by the Sun and always in the same direction, when the tides are produced by it; and in neither case do they tell us how or when the Earth gets back into its natural orbital position again after any given meridian, by the Earth's rotation on its axis, has passed the attractive influence of the Moon or Sun. If there has ever been any recoil detected in the forces which produce the tides, the observers have failed to give a record of it to the public.

As the Moon apparently moves to and from any position at the rate of about thirty-five miles per minute it evidently would cause no sudden movement of the Earth towards it, much less one that would be sufficient to leave its waters behind it. To examine this prevalent theory in regard to the tides, as popularly taught, the objections are more easily explained by reference to the conditions which are claimed to exist and to occur at all the points of the Moon's orbit while the Moon is passing around the Earth, but reference will be made to only a few of the positions of the Moon in which it produces the more prominent results and in which the angles or lines of attraction of the Sun and Moon to the Earth are the least and greatest, leaving the intermediate graduating effect or result for the reader to proportion according to the position that the Moon may be presumed to occupy at the time. During the time of a few successive tides, when the Moon is between or nearly between the Earth and the Sun, the water on the Earth's surface must be drawn toward it quickly to produce the greater tides, and at the same time the Earth must be suddenly drawn, but not quite so far, toward the Moon, in order to leave the water behind it to form the tide which occurs at the same time on the opposite side of the Earth from the Moon, and this same sudden jerking process of the Earth out of its regular orbit, when it is traveling at the rate of about eleven hundred and ten miles per minute, by the Moon which is only about one-fiftieth of its bulk, must occur oftener than once at each revolution of the Earth in relation to the Moon, to produce the several tides which occur on different parts of the Earth's surface while the Earth rotates on its axis once. If the Earth is drawn from its orbit when the tide occurs in the Atlantic ocean, it must be drawn likewise when the tide occurs in the Pacific ocean, and again when it occurs in the Indian ocean. So we see that at three successive times at every rotation of the Earth upon its axis our planet must be drawn suddenly from its orbital position by a force infantile as compared with the Earth, and no system is claimed or advanced by which it may return to its natural orbital position again. When the Moon is at or near the position of first-quarter, it is on or near the Earth's orbit, and directly behind or in the rear of the Earth, while the Earth is traveling at its usual speed right away from the Moon, yet we are asked to believe that the water on the Earth's surface on three separate ocean spaces is drawn back three times at each one of the Earth's rotations, or once for each ocean to each rotation, in order to produce the tides in the different oceans; and also to believe that the Earth is checked suddenly enough in its orbital velocity three times during one rotation on its axis, in order that the water on the opposite side of the Earth from the Moon may apparently pass on a little ahead of the Earth in the direction in which the Earth is traveling, to form the tide on the opposite side of the Earth from the larger one. The effect produced on the Earth by the Moon when it is on the opposite side of the Earth from the Sun, or about the times of being full, is similar to the effect to that which is produced when the Moon is new, or between the Earth and the Sun, it being diminished by the amount of the

Sun's attractive influence on the Earth in causing the solar tide. When the Moon is at or near the last of the third or the beginning of the fourth quarter, it is just in advance of, or in front of, the Earth, with the Earth rushing on toward, or nearly toward, it for several days at its usual speed, and at that time, the Moon must draw the water which is on the side of the Earth toward and nearest it a little faster than the Earth is traveling in its orbit to produce the larger tide, and at the same time it must cause the Earth to suddenly hasten its speed sufficiently to leave a ridge of water on the opposite side, which is shown as the smaller lunar tide, which occurs twelve hours and about twenty-six minutes after the larger one, and these effects must be produced three times to each rotation of the Earth on its axis, in order that one tide may be produced in each of the oceans mentioned, and in which the tides regularly occur. The same inconsistencies appear, no matter in what position the Moon may be to the Earth, when the cause of the tides, as taught in text-books, is sought for in relation to it. So the author feels justified in leaving this theory as he found it, and looking for a cause for the tides that appears to be more in harmony with the operation of the natural forces around us.

The fact that every meridian, or part of the Earth's surface, is being constantly and rapidly presented toward the Sun and Moon, should not be overlooked, for it is in connection and unison with these conditions that the tides occur. The attractive influence of the Moon as causing the tides, is acting upon the Earth in the same manner at all times, and varying in degree as the Moon gradually approaches and leaves its perigee or apogee, and changes its position in relation to the Earth's surface as the Earth rotates on its axis. The effect of the attractive influence on the Earth which causes the tides is always found in the same relative position to the Sun or Moon; consequently there can be no sudden movement to produce conditions which already exist.

Figure V shows the relative position of the tides to the Sun and Moon, as being produced by their attraction and the centrifugal force of the Earth, as developed by the rotation on its axis. The pole of the Earth is shown as being in a horizontal position, with the tides shown a few degrees from and after passing the line of attraction of the Sun and Moon. The solar tide always remains in the same relative position to the Sun. The lunar tide always remains in the same relative position in advance of the Moon, but the rotation of the Earth on its axis carries any given meridian on from the Moon to the position of the lunar tide in the time that the tide appears after the Moon has passed a meridian; this causes the appearance of the tide following the Moon. As the Moon passes around the Earth the lunar tide advances, and passes through, or over, the solar tide, about the time of new moon, or when the Moon passes between the Earth and the Sun, and also when the Earth is between the Moon and the Sun; at these times the solar and lunar tides mingle together and form one which about equals the solar and lunar tides in volume.

The rise and fall of the tide waves seem to add still more proof of the correctness of the theory of centrifugal force, as applied to our Earth and as illustrated by Figure VI. The tides appear to follow the Moon and Sun at certain and regular periods of time, after the Sun or Moon have apparently passed a meridian, and also that they occur again at regular intervals when the Moon or Sun is on the opposite side of the Earth. The small tide that follows the Sun varies in height and volume, according to the distance of the Sun, the same as the lunar tide. If the rapid rotation of the planets on their axes, at any time of their existence, would have caused a greater equatorial than polar diameter, the same law would control, in the same manner, the fluid parts of the planets at the present time; and this seems to be the case in regard to our own planet, whose surface is about three-quarters fluid, and in constant motion, while the surface at the equator is about thirteen miles farther from the center than is the surface of the Earth at the poles.

The wheel, as shown in Figure VI, is the same in principle as the one constructed and tried by the author, which

DIAGRAM, SHOWING THE RELATIVE POSITION OF THE SUN AND MOON TO THE TIDES OF THE EARTH

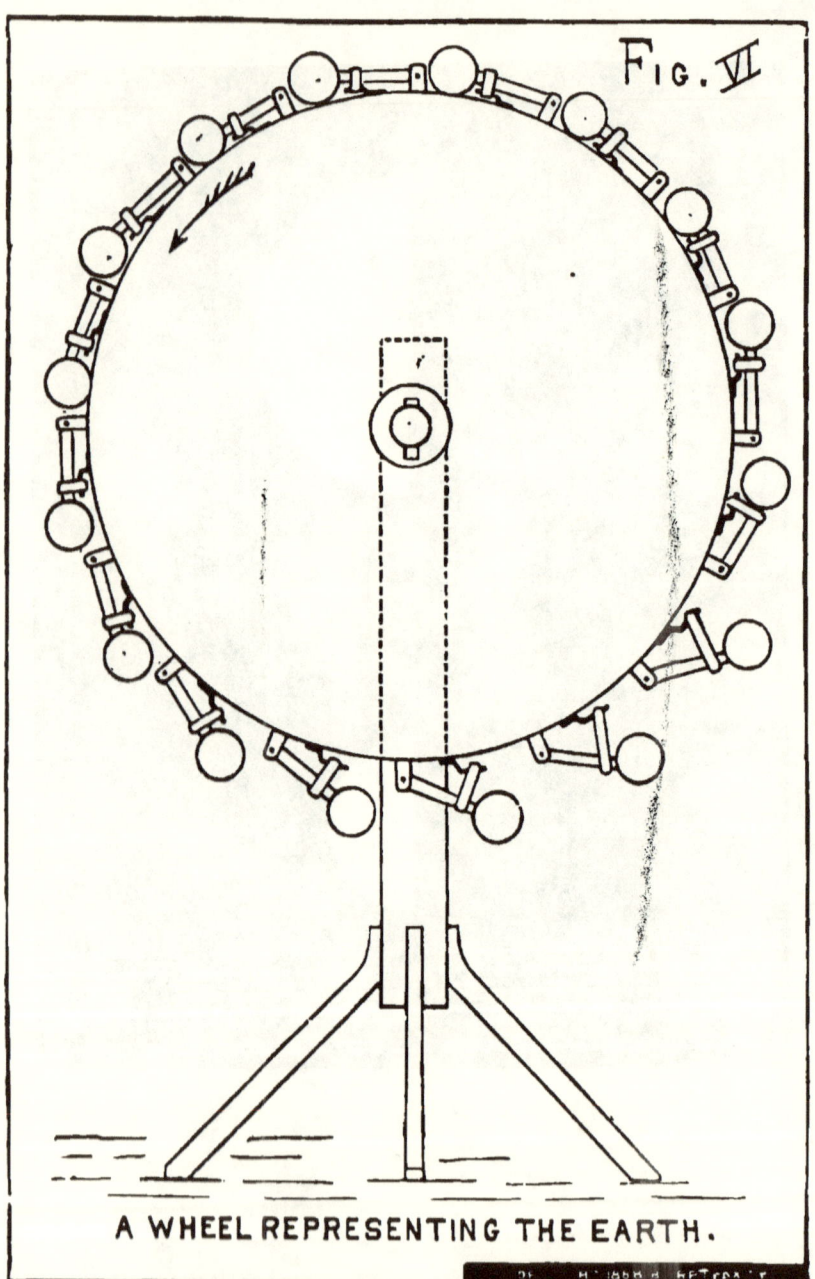

FIG. VI

A WHEEL REPRESENTING THE EARTH.

is intended to illustrate the theory of centrifugal force, as applied to our Earth in causing the tides.

The wheel, as constructed, was about twelve feet in diameter, with levers attached to the rim, with a hinge and weights attached to the end of the levers. The weights were held close to the rim of the wheel by the aid of elastics. The weights were equal in number and weight, so that the wheel was equally balanced, and as attached to the rim of the wheel were intended to represent the water, or any substance that could be moved from its natural place, or position of rest, by the centrifugal force of the Earth, in connection with the attraction of any external body, as the Moon or Sun. The tension on the elastics was equal, and sufficient to hold the weights firmly to the rim of the wheel, when not in motion. The wheel, as a whole, was intended to represent the Earth, with the water of the Earth as being agitated by the attraction of the Moon or Sun, as seen in the tidal action. The elastics, as shown, holding the weights to the rim of the wheel, represent the law of attraction, or gravitation, of the Earth, or centripetal force, as holding the water to the Earth's surface. The Earth's attraction on the wheel represents the attraction of the Moon or Sun on the Earth, only in a far greater degree. When the wheel was placed in a vertical position (or such position as the satellites bear to their primaries, and all of the planets to the Sun), and revolved at a moderate speed, the centrifugal force of the wheel would overcome the centripetal force of the elastics, and the weights would first commence to leave the rim of the wheel soon after they passed the horizontal diameter on the descending side of the wheel, from which point the weights would gradually increase in distance from the rim, until they reached a point about half way between the vertical diameter on the lower side and the horizontal diameter on the ascending side of the wheel. After this latter point was reached, the attraction of the elastics would cause the weights to return quickly to the rim of the wheel, and to remain there in their natural positions until they again passed the horizontal diameter on the descending side, when they would leave the rim of the wheel as before. This greatest throwing-off point corresponds well with the position of high tide on the Earth and the Moon at any given meridian.

In the relative position on the wheel where the low tide should occur on the Earth, twelve hours and about twenty-six minutes after the high tide, or on the opposite side of the Earth from the high tide, there was no perceptible inclination of the weights to leave the rim of the wheel, thereby showing that theories which teach us that the smaller tide is necessary to counter-balance the larger one, that first follows the Moon or Sun, may be erroneous.

The test made with the wheel, as shown in the accompanying figure and as described, would seem to prove that the tidal waves are developed by the centrifugal force of the Earth, whose tendency it is to throw off toward any object which has an attraction for it, and so we have the largest wave occurring at the point indicated, for it is there that the greatest accumulated effect of the Moon's attractive influence is observed. After passing this point on the Earth's surface, or when coming in contact with the shores of continents which are rigid and unyielding, the water is no longer under the influence of the Moon, and then it seeks its natural equilibrium, or level, on the Earth, and in doing so we see the second, or smaller tide, is produced, which may travel a long distance with but little decrease of volume, because of the noncompressibility of water. This theory may also explain why the tides on all eastern shores are higher than on the western shores, and whether or not the wave that approaches the western shores is only a reaction of the one which previously passed to the eastern shore in the same sea, which, if proved to be the case, would account for the tides running from west to east into the bays, harbors, etc., of the western shores of the continents. If the tide waves on the western shores are only the reaction of a greater one going the other way, then (according to this theory), it would be safe to assume that the high tide on all western shores should about equal the one second in height on the eastern shores. By consulting a map of the world or, better still, a globe or sphere with the map of the world upon it, and observing the open space over which the tide wave passes, from east to west, we can get a better idea of the force and direction of the wave, and its terminations against the continents and the final distribution of the water, as shown by the ocean currents, than by almost any other means that we have. It can there be seen that the wave can commence to form along the eastern edge of the Atlantic soon after the Moon has passed the meridian, and appear to follow it, with but little interruption, until it comes in contact with the shores of the American continent, where it will first reach the most eastern point of land, which is at Cape St. Roque, the eastern projection of Brazil, and near the equator. It is here near the equator that the tendency of the Earth is to throw off by its rotation, more than north or south of it, because of its greater circumference in this part. So it follows that the supply that is drawn here must come from parts of the sea which are situated each side of the equator, or from toward the poles of the Earth, and must be drawn toward the equator, as shown by the ocean currents. As the wave in its progress westward is interrupted by the shoals, islands, continents, etc., the sea becomes higher in places than it otherwise would be, and then is forced to seek an equilibrium in the directions where there is the least resistance. One of the most prominent is seen in the warm current of the Gulf Stream that flows northward from near the mouth of the Gulf of Mexico, after it has been driven into the vee (V) shaped form in the Atlantic Ocean, which is formed by the northeast shore of South America, extending from Cape St. Roque to the north of Yucatan, and the southeast shore of North America, from New Foundland to the Florida Reefs. Another, less prominent, is seen in the vee (V) form of the Pacific Ocean, which terminates in the China Sea.

Another simple and easy test can be made by pouring a small quantity of water on a grindstone or similar wheel and revolving it fast enough so that the water will not have time to collect and drop from the lower surface, nor yet fast enough to throw the water off, and conditions will be obtained under which can be seen a quantity of water continually collecting near the point on the edge of the revolving body corresponding with the location on the wheel where the weights were thrown the farthest away, and also corresponding to the position of the tide after any given meridian of the Earth has passed its nearest point of attraction to the Moon or Sun. The author has not taken into consideration the varying times that it takes the tides to run into the various inlets, channels, etc., but has formed his conclusion from effects which have been observed as nearly as possible in the open sea.

This theory, if followed, would also imply that if the speed of the Earth on its axis could be increased sufficiently, the water would be thrown from its surface at nearly the point of high tide, in a somewhat curved line toward the Moon, were it not for the fact that the same law that would throw the water off from the Earth already keeps the Moon at its present distance from the Earth and in its orbit, and an increased velocity of the Earth on its axis would throw the Moon proportionately farther away. The attraction of the Moon would then be lessened in proportion, and the tidal effect would probably be about the same as we see around us at the present time. This same law also implies that an increase or decrease in the speed of the Sun on its axis would enlarge or decrease correspondingly all of the orbits of the planets of our solar system.

### EARTHQUAKES.

It is a natural law in regard to all rotating bodies that the farther from the axis or center of a revolving rotating body or sphere, and the heavier the materials, the greater will be the effect of the centrifugal force in throwing off particles or substances from a rotating body. This would suggest that the higher the continents or land portions, the greater would be the centrifugal effect, so it seems in accord with this law that the Moon should exert a like, if not greater, influence on the higher parts of the Earth than it does on the water, where the effect is so readily and regularly seen. Although the materials of the continents are rigid and immovable, yet it seems that there must be an immense strain upon them every time they are presented by the Earth's rotation to any external body of attraction. A continuous vibration is known to rend the most subtantial structures, if not carefully watched, repaired and protected, therefore it would be difficult to prove that this same law of centrifugal force in connection with the attraction of the Sun and Moon does not aid, to some, if not to a great, extent, in producing our earthquakes. There seems to be no way of ascertaining the time that an earthquake is going to occur, for if they are caused by the centrifugal force of the Earth, there is no way known at present to compute that force, or to ascertain the amount of power that is developed at the Earth's surface, or the amount that is or would be required to cause the earthquakes in the different localities where they have often occured. The fact of their being more frequent in some sections than others may depend somewhat upon the nature of the formation of that locality. The stronger parts of the Earth's surface may withstand the vibrations for hundreds and even thousands of years before the strain will be made known by the earthquake, and after the earthquake has occurred there may be a long or short period of time before the next one occurs, which would probably depend on the solidity and fixed positions of the earthy matter. This strain has not been, nor can it be, easily calculated; for, as in the bridge which sustains a certain weight with safety when new, and which gives way under the same pressure after seasons of vibrations and molecular change, so, after repeated exposures of the continents or land portions of the Earth to the attractive influence of the Moon or Sun, it seems but natural that occasionally there should occur, in some manner, a perceptible result of that power which is applied so often and with such regularity to our planet. In volume XXXI of the *Popular Science Monthly* (page 397), it is stated that M. Perry in his researches, has shown that more earthquakes occur when the Moon is in conjunction and opposition than at intermediate times, and still more when it is directly between the Earth and the Sun, and when the Moon is nearest the Earth than when not in those relative positions to our planet.

On page 398, same volume and same article, Prof. G. H. Darwin, F. R. S., says that earthquakes are more frequent in the winter than in the summer of the northern latitudes. It is also a well-known fact that earthquakes occur oftener within forty-five degrees of the equator than farther north or south of it.

The statements of the above eminent and popular authors seem to fully support the theory of throwing off by centrifugal force, as herein set forth, and as the author has intended to show by the attractive force of the Sun or Moon, or of the Sun and Moon combined, in producing the tides, which vary in height according to the distance of the Sun or Moon. In them we see the effect of the attractive force of the Sun and Moon to be increased or diminished in proportion to the greater or lesser distance that they are from the Earth. As the Sun is somewhat nearer the Earth in our winter than in our summer, so we may expect results that will correspond with the attraction caused by the increased or decreased distance from the Sun, and which would harmonize with the observations of Prof. G. H. Darwin in his researches.

### HEAT.

The question also arises whether much of the heat of the Earth may not be due, to some extent, to the centrifugal force of the Sun acting upon the Earth's surface. Although most authors claim that the heat on the surface of the Earth, whether little or great, is in proportion to the number of the Sun's rays that fall upon each square inch of surface, yet we know that the Earth is several million miles nearer the Sun in our winter than in our summer, and in many locations the land is so formed or shaped that as many, and sometimes more, rays fall to the square inch in winter than in the summer, yet there is much less heat. If the theory that the Earth is slowly cooling down is correct, then it follows that there must be a system or an opportunity by which the internal heat of the Earth may escape or radiate away into the space which surrounds it. The foregoing theory and illustrations are intended to show the manner in which the repelling force of the Sun is acting upon the planets, and the results that are natural to follow such action. The repelling action of the Sun implies a pressure exerted by the Sun upon the bodies upon which it is operating. The greatest heat is where the force rays of the Sun come in contact with the main part of the Earth's surface the more squarely, and it seems but a natural effect that in parts of the Earth near the equator the heat of the Earth would have but little chance to escape, while near the poles, or any part of the Earth which may be so turned from the Sun that its rays will strike it obliquely, the heat of the Earth would meet with little obstruction in escaping, while perhaps the force rays of the Sun in passing by or near the poles might aid in carrying off some of the Earth's heat into space. The snow-capped mountains of the tropics, which are above the suppressed heat of the Earth, are so situated and formed that the heat that is within them can easily radiate or be driven away instead of being driven back or held to the surface of them, as it is in the lower or flat parts of the Earth, where the force rays of the Sun are beating more squarely on the surface of the Earth and holding the heat down, or continually forcing it back to the Earth's surface.

Prof. Tyndall demonstrates in his lectures on heat that there are rays of light and heat separate and distinct from each other, and that, with proper means or arrangements, the light rays can be intercepted and the heat rays, as he terms them, will pass on the same as if the light rays were not shut off, or the heat may be absorbed, and the light rays will pass on as if there had been no interference with the heat rays. The velocity of the sunlight is popularly known to be about 186,000 miles per second, but the author has seen no estimate or calculation of the velocity of the heat rays. Experiments have shown that heat may be developed by submitting substances to pressure or friction. While the attraction of the Sun is drawing the planetary bodies toward it, they are held in position by a repelling force which keeps them a certain distance away from it, and it is not yet proved that there is no heat developed by this repelling force acting on or against planetary surfaces.

Nor is it proved that there is no heat produced by the friction of the Sun's force on entering and passing through our atmosphere. There seems little reason to expect to find heat in the open space of the heavens where the Sun's force rays are not interfered with, unless at or near the Sun's equatorial plane. The sunlight which is reflected from the Moon seems to contain no heat, thus showing that the heat produced by the Sun at or on the Moon's surface must be radiated from or absorbed by it. The attraction of the Earth for the Moon is counterbalanced by the repelling force of the Earth, which is always acting on the same side of the Moon, and if there is any heat developed on the Moon's surface by this repelling force, the temperature on the side of the Moon which is turned toward the Earth may be considerably higher than has generally been supposed. As the Moon travels around the Sun in a slightly scolloped path or orbit, and rotates on its axis only once in about twenty-nine and a half days, it is evident that the throwing off force from it must be very small, and whatever there is must be thrown from its farther side, in part by its apparent swinging motion around the Earth, similar to the water which may be thrown from a sponge when attached to the end of a string and swung in a circle.

Authors have, from time to time, attempted to demonstrate the cause of the Sun's heat, some offering as a reason that the Sun must be gradually shrinking by its own attraction, resulting in the heat which is claimed for it. Some have given estimates of the amount of contraction which would be necessary each year to correspond to the amount of the heat which they have calculated must radiate from the Sun's entire surface in the same time. Some presume the Sun to be surrounded by aerolites and meteoric substances, etc., which are constantly falling to the Sun from every direction in countless millions, thus producing the Sun's heat by their velocities and impact, combustion, etc., on the Sun's surface, all combining to produce the heat of the Sun, which is said to be radiating equally from every part of its surface and in the same direction as the sunlight. Some authors have undertaken to estimate the entire amount of the Sun's heat by assuming that the same amount is being radiated alike from every part of its surface, and they base their estimates on what is received in certain locations on the Earth's surface. These estimates are, probably, a great deal too much, if the theory that the Sun is throwing off influences by centrifugal force is correct. For heat, unlike light rays, can be changed from a direct course, or line, by and in the direction of a passing current, as shown by the heat of the simoons of the desert, or the moderated temperature in countries of the northern hemisphere, caused by the flow of ocean currents from the tropic regions, or as may be shown by the heat which may be driven off with the air in any direction from a heated body.

The author admits that the heat may be radiating from the Sun's entire surface, in nearly the same amount, but claims that after it radiates a certain distance from its surface it is met by the inflow of the influences which are being constantly drawn toward the Sun and, from there, most of the heat is diverged or drawn toward the Sun's equatorial region, to be carried off in the current of the Sun's force, as shown in the sectional edge-view of the Sun in Figure II. In this manner, most of the heat that is radiated from its surface finds its way into the force current and is distributed in the region of the Sun's equatorial plane. Thus we see that the estimates of the Sun's heat must be entirely too great, when calculations are based on the amount indicated in different places on the Earth's surface, because the Earth is always in or near the line of the Sun's equator.

The author is also willing to admit that many aerolites, etc., may be falling to the Sun's surface, but he claims that they must fall to the surface only at those parts where the influences are approaching the northerly or southerly surfaces, and those that do reach the Sun's surface, in all probability, remain there, the same as they are known to remain on the Earth after once reaching it, unless converted into gases by the heat of the Sun, in which case they may be driven off again. It is neither claimed nor admitted that a combustion of the elements or influences which are drawn to and sent away from the Sun is necessary to maintain life and stimulate growth on the Earth.

The writer is also of the opinion that but little of the influences which are drawn toward the Sun, at the sides of the equatorial region, comes in contact with the Sun, but that they may be drawn into the stream of the Sun's force and carried away many times before reaching the Sun's surface.

It seems that if this force were thrown off before having an opportunity to reach the Sun's surface, it would be just as capable of carrying the heat of the Sun as if it had been in contact with its surface. Much of the heat may be due to the friction and agitation of the influences while they are being drawn near to and thrown away from the Sun by its centrifugal force. In either case the heat seems to be carried by the influences which are thrown from the Sun. These influences appear to be drawn to the Sun at all times alike. A greater or less quantity of them, at times may account for the changes occurring on the Sun's surface.

There seems to be no system by which the Sun's force approaches the Sun that would imply that the influences are always attracted in a like amount, or in exact or proportional elements. It appears to attract all substances to itself in the same way, and may, at times and places, show by a slight difference in its light a different composition of the elements which are drawn to it, as in the case of some comets, which attract substances to themselves from the space through which they are traveling, which do not, on their return, indicate the same composition as on former visits.

All of the rotating planets may be presumed to throw off their heat in the same manner as the Sun and, in amount and degree, according to the temperature of each individual planet.

## PLANETARY FORMATION.

The two forces, attraction of gravitation and repulsion as heretofore represented, appear to have been all that were in operation in forming the Sun and planetary bodies. Although the substance-matter and the laws of attraction of gravitation and repulsion seem infinite and co-existent, and without beginning, yet the effect of attraction on the substance-matter must have preceded, in effect, the repelling action for a long period of time; for a substance must assume form, then axial rotation, in order to develop the repelling force. In the case of our Sun, the greater portion of the substances of which it is composed may have been drawn together before it acquired an axial velocity that was sufficient to effectually repel the substances of which the planets are formed, for we see that the several combined amounts of the planets and satellites do not equal one five-hundredth part of the Sun. The force of the Sun, like any other force, is capable of accomplishing a certain amount, and no more. That is, it will repel substances until the force has become disseminated and weakened, when the Sun's attraction will prevent the substance from going farther. These forces of the Sun seem to have located, or retained, the planetary substance-matter in their earlier or original locations, in belts or zones around the Sun, at distances at which the forces of the Sun could suspend it, according to the gravities of the substances, after which nuclei, or centers of attraction, were formed by the Sun's agitation, then axial rotation and revolution around the Sun, the larger body attracting to itself the smaller amounts, from a distance each side as far as its attraction could control the substances in opposition to the Sun's force, beyond which the substance-matter seeks the next center of attraction. These same conditions apply to the formation of the satellites by the attractive and repelling forces of the primaries.

The rings of Saturn seem to furnish a good illustration of suspended planetary substance, but if the rings are revolving with, and at the same time as, the planet, as some observers claim, the agitation of their substances must be

www.ingramcontent.com/pod-product-compliance
Lightning Source LLC
Chambersburg PA
CBHW031245260626
47169CB00007B/2452